THE DOWNSIDE OF AI

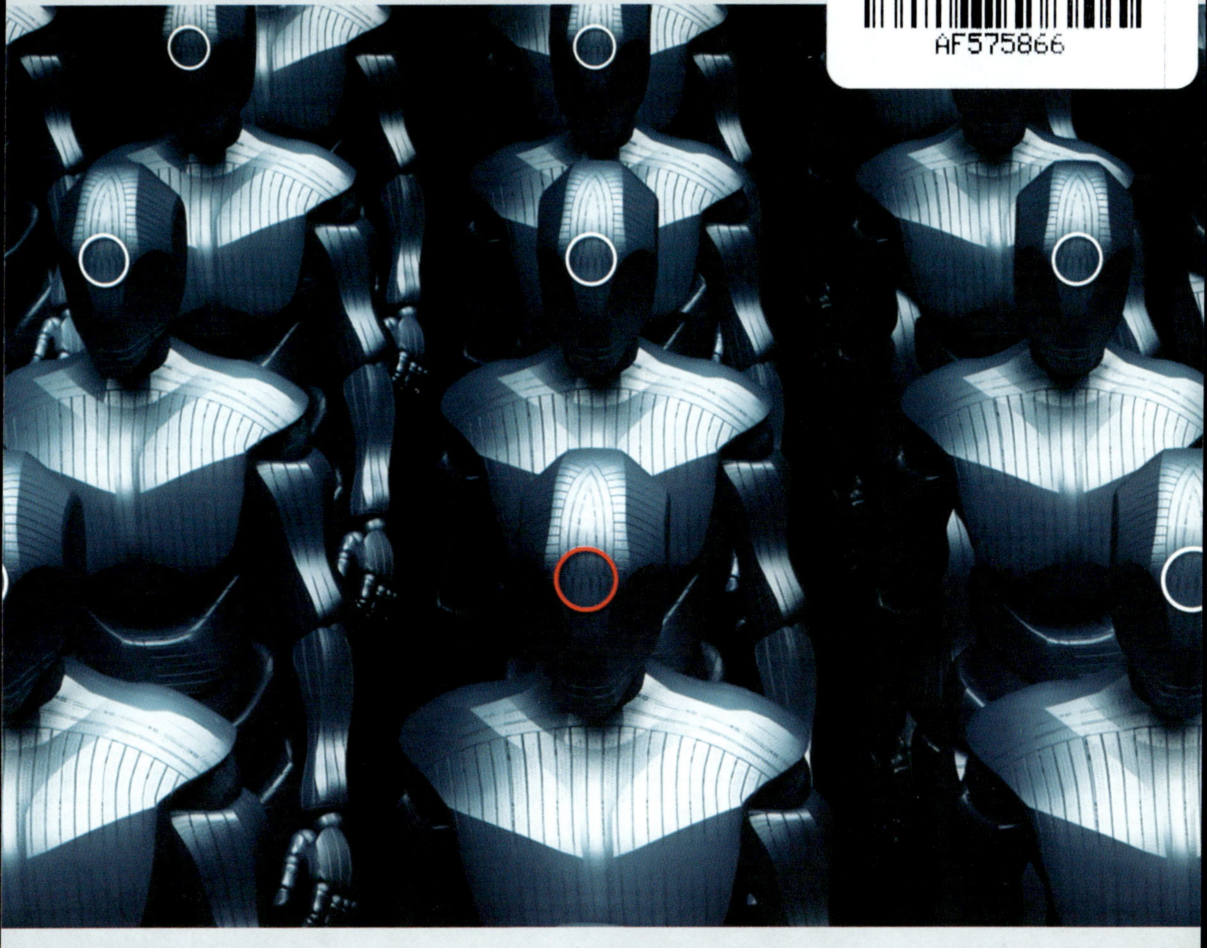

by Josh Gregory

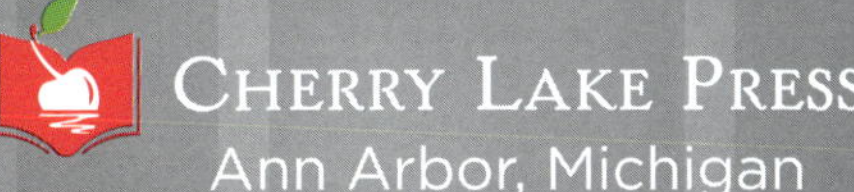

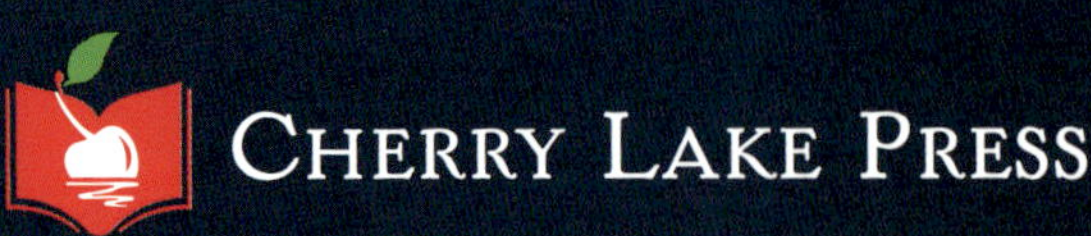

Published in the United States of America by Cherry Lake Publishing
Ann Arbor, Michigan
www.cherrylakepublishing.com

Reading Adviser: Beth Walker Gambro, MS, Ed., Reading Consultant, Yorkville, IL

Photo Credits: © yucelyilmaz/Shutterstock, cover, title page; © jittawit21/Shutterstock, 5; Prototyperspective, CC0, via Wikimedia Commons, 6; The White House, Public domain, via Wikimedia Commons, 9; © Phonlamai Photo/Shutterstock, 11; © fizkes/Shutterstock, 13; © Stock-Asso/Shutterstock, 17; © wavebreakmedia/Shutterstock, 19; © Markus Mainka/Shutterstock, 21; © Flash Vector/Shutterstock, 22; © Gorodenkoff/Shutterstock, 25; © Suri_Studio/Shutterstock, 27; © Kittipong Jirasukhanont/Dreamstime.com, 28

Cherry Lake Press is an imprint of Cherry Lake Publishing Group.

Library of Congress Cataloging-in-Publication Data has been filed and is available at catalog.loc.gov.

Cherry Lake Press would like to acknowledge the work of the Partnership for 21st Century Learning, a Network of Battelle for Kids. Please visit Battelle for Kids online for more information.

Printed in the United States of America

ABOUT THE AUTHOR

Josh Gregory is the author of more than 200 books for kids. He has written about everything from animals to technology to history. A graduate of the University of Missouri–Columbia, he currently lives in Chicago, Illinois.

CONTENTS

Chapter 1

TECH TROUBLE

Artificial intelligence (AI) has taken the world by storm. Companies are finding new uses for this incredible technology. Inventors are making it even more powerful. Everyone is talking about it. You have probably heard about it online or in the news.

AI is a type of computer program. But it is different from other kinds of programs. Other programs simply follow the exact instructions in their **code**. They can only get better if **developers** write new code. AI is different. It is a computer program that is meant to work like a human mind. It looks at **data**. It finds patterns. Looking at more data makes it better at things. It can learn and improve. This is a lot like how practicing can help you get better at a musical instrument or playing a sport.

You might have noticed that some people are very excited about AI. They think it will make the world a better place. AI is already helping in many ways.

Many people believe that using AI will improve our daily lives. They believe we will learn how to work well with AI.

Researchers are using it to help create new kinds of medicine. Doctors are using it to help perform surgeries. Teachers are using it in classrooms. Companies are finding ways to make more money using AI. You probably use AI technology every day without knowing it. AI recommends you new videos on YouTube. It decides which posts you will see on social media. It helps decide what websites you'll see when you Google something.

A lot of people disagree about the future of AI. They think AI is going to cause big problems. Some argue that AI will help only certain people. These people will get richer and live better lives. Others will lose their jobs, or worse. AI's biggest critics even argue that AI could one day cause the end of our society.

Some people think there will be an AI uprising. They think this will be the end of human society.

Some critics are people who don't understand how AI works. But others are experts in the field. They know very well what AI can do. And they want to warn people about possible dangers.

One possible problem with AI is that it could make people lazy. AI would make all their decisions. It would replace many kinds of jobs. People might forget how to do things for themselves. They would no longer be able to live without AI technology.

Even the most powerful AI is still a computer program. It can do only what it is programmed to do. AI critics point out that AI does not have emotions. It doesn't think about the way its decisions will make people feel. It doesn't care whether its actions will cause problems for humans. Putting AI in charge of important tasks could be very **efficient**. But that doesn't mean it will be good for people.

The debate over AI isn't going anywhere. It will only get more complicated as new AI technology is created. People will always have different views of AI. But we will all need to agree on the basics. How big a role will AI play in society? Should we use it only for certain things? Do we need to be careful with it? There is a lot to think about. Knowing more will let us make better decisions.

THE LAW OF THE LAND

One way to deal with AI might be to pass new laws. These laws could keep AI from being used in certain ways. Or they could protect people from the worst effects of AI. AI critics argue that governments should act now to create these laws. They believe that it will be too late if AI gets more powerful.

Others argue that laws will prevent AI technology from reaching its potential. They believe that inventors should have freedom to use AI however they can. This could lead to useful new forms of AI. But it could also lead to powerful and harmful uses.

President Biden and Vice President Harris met with CEOs of businesses that use AI technology. Lawmakers at the federal level are learning what laws may be needed to protect citizens from possible downfalls of AI.

Chapter 2

CHANGING THE WAY WE WORK

One of the biggest fears about AI is that it will take jobs from human workers. AI is already being used to **automate** many jobs. AI can operate robots in factories. It can answer customers' questions in online chats. Some companies have replaced writers with text-generating AI programs like ChatGPT. McDonald's has even opened automated restaurants. Customers can order food through an app. Robots serve the food. The customers never meet a human worker.

Many people fear that AI will one day replace most jobs. This will make it harder for people to find work. They won't be able to make money. People could go hungry. They could lose their homes.

Some jobs might not be fully automated. But these workers may still have to use AI. A company might

Many big car companies currently use robots in their manufacturing factories.

replace five workers with one worker who oversees an AI system. This saves the company money. But it comes at the cost of jobs for humans.

All AI programs are created by humans. Humans can make mistakes. This means AI can be flawed. It doesn't always work the way people want it to. This can cause big problems in some cases. Some companies are using AI to help hire new employees. AI can sort through job applications. It can pick the applicants it thinks will be the best fit. But what if it isn't looking for the right things? And what if it isn't fair? Some AI systems have shown racial **bias**. The data used to train them was flawed. Imagine this happened with an AI used to hire people. It might make choices based on the color of an applicant's skin. This would be racist and unfair.

AI isn't just changing the way people work. It is also changing the way they learn. Teachers are finding interesting ways to use AI. It helps them design lessons that work for each student. And one day, students might have AI tutors that answer questions about homework.

AI can be a good tool for learning. But some people are concerned that students will use it to cheat. Students might use AI programs like ChatGPT to write essays. It can be hard for teachers to know when this is happening. Students might get good grades without learning anything. They won't have the skills they need to find good jobs after they finish school.

Some students may use AI programs to write essays. But AI programs also exist to help teachers check for the use of AI in student papers.

ASK AN AI

Let's go straight to the source and ask the AI text generator ChatGPT about job automation.

Hi, ChatGPT
Hello! How can I assist you today?

How is AI automating jobs?
AI automates jobs by using robots and chatbots to do repetitive tasks faster, like assembling things or answering questions. This can make things more efficient, but some worry it might replace human jobs.

Will there still be jobs for humans in the future?
Yes, there may be changes in the types of jobs, but humans will likely still have important roles, especially in creative and critical thinking tasks that AI can't do.

How can you be sure?
We can't be certain about the future, but in the past, when technology changed, new job opportunities appeared. As AI advances, there will likely be a need for skills like problem-solving and communication that humans excel in. Being adaptable and learning new skills will be important for staying relevant in the job market.

Thanks for sharing your thoughts!
You're welcome! If you have any more questions or if there's anything else I can help you with, feel free to ask.

(This conversation has been slightly edited and condensed for space.)

Chapter 3

CREATIVE CRITICISM

Many of the most exciting new uses for AI involve creativity. AI programs can create realistic photos and videos. They can write text about any topic. They can even make music. And they are very easy to use. All you need to do is type what you want the AI to make. It will then give you what it thinks you want.

AI-created content is impressive. But critics argue that it is not as good as content that people make. Art comes from humans expressing their feelings and ideas. AI programs do not have feelings or ideas. They cannot express themselves. All they can do is imitate what people have already done. This means that many people do not consider AI-created content to be real art.

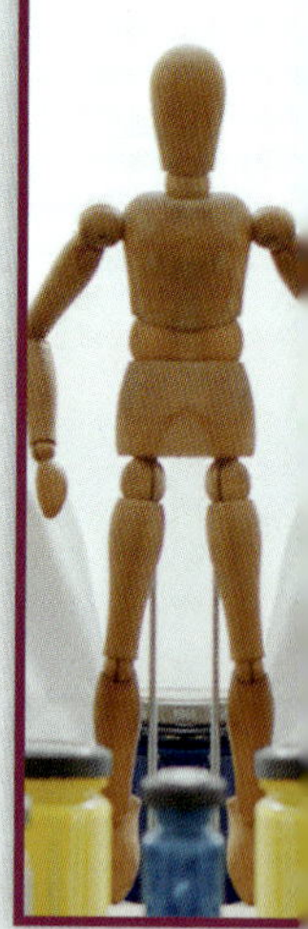

Even if AI is able to create art, AI doesn't have feelings or ideas to express. Many people argue that this makes human art better.

Many artists are still concerned that AI will take away their job opportunities, though. Companies will think AI-created content will be "good enough." They will want to save money. So they won't pay human artists to create the best art possible.

Similarly, some are concerned that AI will lower people's standards for art. People might get used to hearing music made by AI. Or they might get used to reading books written by AI. They would forget how much better real human-created music, artwork, and books can be.

There are many people that believe AI will never be able to replace human artists.

Even the most impressive AI creations are not the same as human-created art. AI programs cannot create their own original ideas. They create things by combining ideas from human-created artwork. AI developers show lots of human-created art to their AI programs as part of the training process. For example, ChatGPT was trained using books, magazines, and online posts written by humans.

4:37
chat.openai.com
20
New chat
ChatGPT
Examples

AI programs look for patterns in human-created art. Then they use this information to "create" new things. But these things can only come from other people's ideas. Many artists argue that this is a type of **plagiarism**. The AI programs are stealing the original work of others. Some have sued AI companies. They believe their work should be protected by law. They don't want AI programs to use their ideas.

THE DEEPFAKE DILEMMA

AI is getting very good at creating things like photos and videos. Sometimes these images look almost real. But this can cause problems. Deepfakes are videos that look and sound real. They often show celebrities and other real people. But they aren't real. They can make it look like real people are saying or doing embarrassing things. This is bad for the people shown in the deepfake videos. But it is also bad for everyone else. People will have a harder time knowing what is and isn't real. They won't trust the things they see in videos and photos.

Chapter 4

OUT OF CONTROL

AI has become very powerful in just a few years. Its abilities are growing faster and faster. It is being used for more and more things. Some people wonder if we are moving too quickly. Could AI become too powerful in the coming years?

Many people are worried that AI will ruin their privacy. AI can identify people in photos and videos. Sometimes this is no big deal. You post a photo of your friends to social media. AI recognizes all of your friends and tags them. This is convenient. But there are cameras everywhere. Some people record videos of strangers in public. Phones and other devices also track where people go.

AI can use this information to predict people's schedules. This might not sound like a big problem at first. But some people do not want to be tracked or recorded. They want to have privacy. The spread of AI might one day make that impossible.

It may become more and more difficult to maintain privacy as AI becomes more advanced.

AI technology is making its way into many important systems. Banks use it. Utility companies use it. We rely on these things every day. But we know that AI isn't perfect. Some people worry that it could cause problems in these important systems. And that would cause problems in everyone's lives.

Soon, wars could be fought by AI-powered weapons. Militaries are already using AI-powered **drones** to gather information. And human-controlled drones are used to launch attacks. These tools could be combined. AI-powered drones could be given weapons. But what if these AI programs make mistakes? They could injure or kill innocent people.

People are discussing ways to regulate AI. Decisions made now will influence how AI is used in the future.

It's too early to know exactly how AI will change our world. We can't predict everything that will happen. But the decisions we make now will have a big effect. Will we let this technology take over everything we do? Or will we be careful about how we use it? AI could help people in many ways. But it could also cause big problems. It is important to think carefully about how we use it.

Some people think AI could take over society, like in a sci-fi book or movie. That is not currently possible.

A MIND OF ITS OWN

Today's AI cannot think for itself. It can only do what humans program it to do. But some people fear that AI could one day learn to think for itself. It could be able to outsmart humans. It could also create its own new AI programs. This would let AI fight against humans. It could take over the world.

This sounds like the plot of a science-fiction movie. Right now it is completely impossible. Some people believe it will never be possible. But many types of technology once seemed impossible. Some argue that we should already be working to make sure AI can never learn to think for itself. What do you think?

ACTIVITY: MAKE YOUR OWN RULES

Imagine you are a government leader. Would AI be a big issue for you? Try writing your own law to deal with the problems raised by AI. Here are some things to think about as you write:

- What problem do you want to solve? AI is a big issue. It deals with many different things. Pick one thing you'd like to solve. You can look back through this book for examples. Or maybe you'd like to pass a law in support of AI. It's up to you.
- How will you solve the problem? Write down all of the steps. Be as detailed as you can.
- Could your law introduce new problems? Think carefully. Try to think about the consequences of your law.
- Do you think people would support your law? Try asking friends and family what they think.

Making new laws isn't easy. There are many details to consider. But good laws can help improve life for everyone!

FIND OUT MORE

Books

Abell, Tracy. *Artificial Intelligence Ethics and Debates.* Lake Elmo, MN: Focus Readers, 2020.

Felix, Rebecca. *Artificial Intelligence: Can Computers Take Over?* Minneapolis, MN: Checkerboard Library, 2019.

Gregory, Josh. *Careers in Artificial Intelligence.* Ann Arbor, MI: Cherry Lake Publishing, 2019.

Kulz, George Anthony. *Artificial Intelligence in the Real World.* Lake Elmo, MN: Focus Readers, 2020.

On the Web

Search these online sources with an adult:

"Artificial Intelligence." Britannica for Kids.

"Artificial intelligence facts for kids." Kiddle.

"ChatGPT." OpenAI.

"What is artificial intelligence (AI)?" IBM.

"Will Future Robots and A.I. Take Over?" National Geographic.

GLOSSARY

automate (AW-tuh-mayt)
replace human workers with computer systems

bias (BYE-uhs)
an unfair preference for something

code (KOHD)
instructions written in computer programming language

data (DAY-tuh)
information used to create, process, or support something

developers (dih-VEH-luh-purz)
people who create AI and other computer programs

drones (DROHNZ)
remote-controlled aircraft or ships

efficient (ih-FIH-shunt)
able to do something using as few resources as possible, such as time

plagiarism (PLAY-juh-rih-zuhm)
the act of stealing others' work or ideas and passing them off as your own

INDEX